WILTSHIRE WINDMILLS

Martin Watts

Wiltshire Library & Museum Service

1980

VISITING WILTSHIRE WINDMILLS AND SITES

It should be noted that, with the exception of Wilton Mill, Grafton which is regularly open to the public, the majority of Wiltshire windmill sites and remains are on private ground and access at any site should not be assumed without consent.

Full six-figure National Grid references have been given wherever possible to enable the sites and remains to be located, but private ownership and the country code should be respected in all cases

Front cover:
Wilton Mill, Grafton (WCC).
Title page:
Devizes. The windmills on the castle mound, from Stukeley's drawing of 1723.
Opposite:
Broadchalke. Detail from the 15th century wall painting, now destroyed (WRO).

© MARTIN WATTS, 1980
Published by Wiltshire County Council Library & Museum Service
Director: Frederick Hallworth, OBE, FLA, FRGS
Edited and designed by Edward J. Kelly, ALA
Printed by Dawson & Goodall Ltd, The Mendip Press, Parsonage Lane, Bath
ISBN 0 86080 067 9

CONTENTS

Wiltshire windmills, a brief history   7

Some medieval references to windmills   15

Windmills and sites listed geographically   17

Tower mill remains   27

Wilton windmill, Grafton   35

Notes and references   43

ACKNOWLEDGEMENTS

In any work of historical research relating to a specific county, the author's main debt must be to the wealth of material in the various archives and its access through archivists and librarians. I am indebted to all the staff of the Wiltshire Record Office in Trowbridge (WRO) for their help, in particular to Ken Rogers who has helped with numerous references found through his own researches and made freely available. I am also grateful to John Sawtell and Angela Harrington; Paul Robinson, and the staff of Devizes Museum; the late Richard Sandell, former librarian of the Wiltshire Archaeological & Natural History Society (WA&NHS), Devizes; Mike Corfield, and the Library & Museum Service of Wiltshire County Council; Mrs Margaret Waley of Great Cheverell; Peter Nicholson of Salisbury and Don Cross of Shrewton. Outside Wiltshire, Mrs Monica Dance, formerly secretary of the Society for the Protection of Ancient Buildings (SPAB), allowed me unlimited access to the Wind & Watermill Section's library and Ken Major has made available material from his collection relating to Wiltshire mills. The Science Museum Library has allowed me to make use of the valuable material in the collection of the late H. E. S. Simmons (SC), and I am pleased to acknowledge the help of many of my 'mill' friends, in particular Geoff Bridger, Frank Gregory, Max Hoather and David Nicholls of Messrs Jameson Marshall Ltd., who have all assisted with information and photographs; also the many tolerant and friendly people who have allowed me access to sites and have tried to answer my questions, in person or through correspondence. Finally, but not least, my thanks to Jill, my wife, for her patience and interest at all times.

Boscombe post mill, Allington (left) and Imber tower mill, from tithe maps (WRO).

ILLUSTRATIONS

Wilton Mill . . . . Front cover
Map of mill sites . . . . Inside front cover
Mills at:
    Devizes . . . . Title page
    Broadchalke 3
    Boscombe 4
    Allington 4
    Brinkworth 6
    East Knoyle 10
    Winterslow 13
    Kemble. 14
    Tilshead 14
    Aldbourne 16
    Chute 18
    Bradford on Avon 25
    Brinkworth 26
    Chiseldon 28
    Brinkworth 28
    Devizes. 30
    Idmiston . . . . .32 & 33
    Wilton. 34, 36, 38, 40, 41, 42
    Winterbourne Monkton 47

All photographs are by the author unless acknowledged in the captions

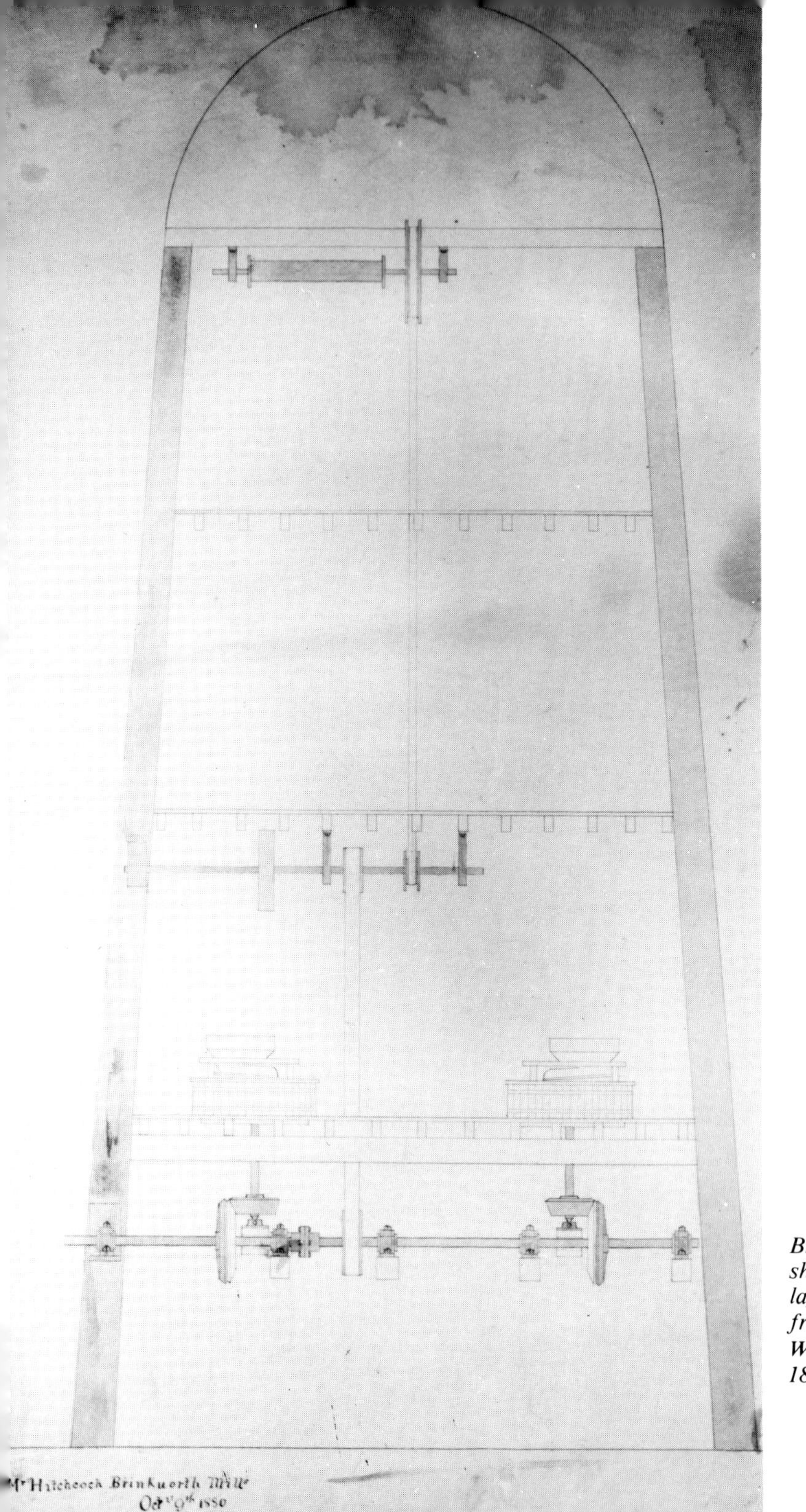

Brinkworth Mill, showing two pair layout, a drawing from Westport Iron Works, Malmesbury, 1880. (WCC).

Wiltshire Windmills, a Brief History

With the recent restoration of Wilton Mill, Grafton, to full working order, the subject of windmills in the county of Wiltshire is deserving of some notice, to put Wilton Mill in some sort of historical and technical context.

The agricultural tradition of Wiltshire was a finely balanced sheep and corn husbandry in the central chalkland area, with cattle and dairying dominant in the clay vales to the north, making best use of the natural make-up of the county. The natural resources were further exploited in terms of watercourses used to power mills. At the time of the Domesday Book, 1086, there were mills at 197 of 335 settlements in Wiltshire, the greatest density being in the Wylye valley.[1] The water-powered corn mill was established in England before the Norman conquest and by the end of the 12th century water power was also used to drive fulling mills, for the finishing of woollen cloth. In Wiltshire the woollen industry reached its peak early in the 19th century, with watermills and factories on most of the major streams, particularly in the west.[2] Alongside the cloth mills, often literally, were the corn mills, always an important part of local economy, reducing cereal grains to flour and meal for human and animal consumption. In areas where corn was grown but there was little potential for water-power, on the chalk downs for instance, windmills were built for corn grinding. In a county with several substantial watercourses, watermills have tended to be the natural preference and, once a site was established with the necessary and often costly dams, weirs and ponds, these mills became sound investments for rebuilding and modernisation. Windmills were always subject to natural wastage through storm and fire damage, and at any site where long occupation is claimed the mill would have been rebuilt and repaired frequently.

The first authentic records of windmills in western Europe appear during the last quarter of the 12th century and by the mid 13th century references exist to windmills throughout England. The earliest so far known in Wiltshire is to a windmill valued one mark on the manor of Hebelesburnel, identified as Ebbesbourne Wake in the south of the county, in 1248.[3] The need for motive power in the medieval period to drive corn mills is underlined by the number of references to water, wind and horse mills, a need created by the demands of a cereal-based staple diet and an increasing population and enforced by the manorial soke, the tenants' obligation to have their grain milled at the lord's mill, with a toll taken in kind.

A *"molendinum ventricum"* is referred to at Highworth in 1275 and a *"molendino ad ventum de novo faciendo"* — a windmill newly built — at a cost of £12. 2s. 10d., at Sevenhampton in c.1285. Both mills were on the lands of Adam de Stratton and when his estate was taken into the king's hands in 1289 it included a watermill in a ruined state which was subsequently repaired and two millstones were chosen at Tewkesbury and 'carried thence' at a total cost of 37s., one for the watermill and the other for the Highworth windmill.[4] Millstones represented quite a considerable capital outlay, particularly in Wiltshire where there is no natural source and they needed to be imported.

At Sherston a windmill valued at 2s., a horse mill worth 10s. and three watermills worth 36s. were held by the manor in 1375. The windmill and horse mill supplemented the available water-power, for in 1349 it was noted that two of the manor's watermills were worth 20s. a year "and not more because they cannot grind in summertime".[5] The frequent occurrence of the place-name 'Winterbourne' in Wiltshire serves as a reminder of the seasonal nature of many of the county's watercourses and, with the vagaries of the wind also, the output of these medieval mills must have been limited.

Evidence of early windmills and sites is scarce but even where no mill has stood for perhaps 500 years the name often survives, 'Windmill Hill' being the most common. Such sites can be traced from the Tithe Awards and large scale maps or located in the field in the form of mounds, often formerly considered to be round barrows. A tumulus in Redlynch parish, for example, marked on the Ordnance Survey maps (SU 202 218) is probably a windmill site, standing as it does in a field called 'Windmills Piece' in the Tithe Award. Similarly at Whiteparish two mounds survive, the sites of windmills built for two different manors in the late middle ages. In this case a general population expansion was reflected in an extension of arable land which, in turn, resulted in the need for more mills.[6]

Changes in manorial control after the dissolution of the monasteries, c.1540, undoubtedly resulted in further windmill losses. A ruined windmill on the land of Thomas Chaderton in Lydiard Millicent in 1586[7] was perhaps the victim of a change in land use or simply surplus to the new owner's requirements and so allowed to run down or be left unrepaired after storm damage. No later reference to this mill is known.

The earliest depiction of a windmill in Wiltshire was perhaps the 15th century wall painting in All Saints Church, Broadchalke, destroyed during a 'restoration' of 1847, where a small tower mill was shown over the right shoulder of St. Christopher.[8] Similar paintings survive, though restored, in Somerset churches, and the association of windmills with the saint of travellers is probably due to the good vantage points which made windmills prominent landmarks. The windmills marked on Saxton's county map of 1576[9] are obvious landmarks and the fact that they are only shown at Clyffe Pypard, where a windmill existed as early as 1361[10] and a Windmill Hill still exists, on South Down, Ebbesbourne Wake, and at Tidworth, does not mean they were the only mills standing at that time, but all would have been valuable landmarks. Similarly John Ogilby's road maps of c.1675 show windmills close to the main roads and again only three are indicated in Wiltshire.[11] One is at Shrewton, one on Windmill Hill, Tidpit, in the parish of Martin which was transferred to Hampshire in 1895, and the third at Kemble, a parish which became part of Gloucestershire in 1896.

Evidence exists at present of 102 windmills in 74 parishes, over a period of some 650 years, but it is not until the late 18th century that sources are available which enable actual sites and details of the mills which occupied them to be recorded accurately. Andrews and Dury's fine county map of 1773[12] shows 13 windmills and indicates a further 8 sites by place names. The little windmill symbols on this map also set out to define post mills and tower mills. The 2″ to 1 mile preparation maps made by the Ordnance Survey between 1808 and 1830,[13] from which the 1st edition of the 1″ to 1 mile maps were compiled, are a valuable source of mill sites and Greenwood's map of 1820 locates 20 windmills and 4 other sites.[14] The turn of the 18th century seems to have been the hey-day of windmill building in Wiltshire, the result of a number of social and economic pressures. The population was rising generally as a result of the so-called 'industrial revolution' and the effects of the Napoleonic wars were felt in most of the southern counties, in terms of man-power available for agriculture and emphasis on local corn growing. Some windmills had relatively short working lives because of the period of depression which followed the wars, but many worked on into the second half of the 19th century.

The 19th century windmills of Wiltshire represent an interesting turning point in English windmill technology, coming between the well-found mills of eastern England and the home counties, with their merchant trade influenced by their nearness to London, the millers being involved in buying in grain in bulk and milling and retailing the products, and the localised and mechanically unsophisticated tower mills of Somerset, each serving a small community and grinding each farmer's grain separately with the miller extracting his toll in kind after the medieval tradition.[15] This local trade came under great pressure after the introduction of free trade in the mid 19th century, when hard imported wheats were ground in purpose-built mills at the port of entry and the

widespread use of steam power, both in transport and agriculture, changed the pattern of the Wiltshire countryside. Many mills became redundant; those away from large centres of population and railways survived by meeting local needs, but windmills were not always able to compete with the more reliable watermills and steam mills and there was a swing away from the finely balanced sheep and corn economy to mechanised farming and increased dairy and meat production, with the produce easily transported by rail to London.[16] By the end of the 19th century those Wiltshire windmills still at work were reduced mainly to a gristing or provender trade, supplying animal feeds, and few survived the turn of the century. Wilton Mill was the last to work by wind alone, closing down just before the start of World War I.

The majority of Wiltshire windmills were corn mills, grinding mainly wheat and barley and most mills had at least two pairs of stones. French burrs were used for wheat milling, grinding fine flour for human consumption, and a pair of grist stones — usually Derbyshire peaks or greys or Welsh conglomerates — were used for provender milling, grinding mainly barley for animal feed. Many mills also had bolters or dressers for cleaning and grading the meal and the windmill at Collingbourne Kingston had a stable by "with a kiln for drying wheat only" in 1839.[17] An unprovenanced windmill, to let in 1857 by Mr Hunt, millwright, of Calne, contained one pair of French and one pair of Welsh stones, two flour machines and a smutter, and there was a six quarter malthouse and beerhouse nearby.[18]

Two tower mills were built in Devizes in the early 18th century for grinding rape seed for the production of a vegetable oil, then much in demand for lighting and lubrication purposes, but by the 1740s they were used, as were several watermills in the Devizes area, for grinding snuff.

Water-raising machinery, important particularly on the edges of the chalk downs where the wells are deep, was initially worked by hand or animal power, but by the end of the 19th century wind pumps were in use on many farms and estates. John Wallis Titt, the son of a windmiller at Chitterne, was trained by the Devizes engineers Brown & May and later set up his own agricultural implement works at Warminster where water supply equipment and wind engines, annular-sailed wind pumps, were produced. The firm is still in operation, but few of the wind pumps they built survive intact.[19]

Traditional mill building was a specialised craft, combining the skills of carpentry, wheelwrighting, masonry and smithing and the name 'millwright' for a maker of mills is known in Wiltshire from at least the beginning of the 17th century.[20] Wind and watermills were virtually the only large pieces of complex production machinery at work up to the end of the 18th century and each area needed several millwrights to build and maintain them. The millers themselves were of necessity practical men and would have tackled many of the day to day jobs of maintenance.

Up to the end of the 18th century the majority of millwork was of timber, with iron fastenings and iron journals and bearing surfaces let in. In 1746 James Burrough, bell founder at the Mint, Devizes, advertised the supply of "brasses for wind or water mills"[21] and in 1799 William Watts, millwright of Great Cheverell, advertised the leases of two windmills on Salisbury Plain and that "any person desirous of erecting a windmill may be accommodated with the materials and workmanship completely fitted for the purpose, which may be erected so as to proceed to work the mill within 3 months".[22] Great Cheverell was later the home of one of Wiltshire's best represented millwrighting families, the Dunfords, whose work may still be found in several of the county's watermills.[23] By the 1820s cast iron for shafting and gearing was becoming common and Wilton Mill, Grafton, was built in 1821 "by an eminant (sic) millwright upon a most modern and improved principle; and nearly the whole of the machinery is cast iron".[24]

While the materials of the mill buildings themselves tend to follow the local traditions, brick in the east, brick and flint in the south east, timber framing in the central/Plain area and stone in the north, west and south west, the windmills to the east of the county were also built with several of the technical developments of the 18th and 19th centuries, such as shuttered sails, fantails and cast iron shafting and gearing. Wilton Mill and West Gomeldon Mill, Idmiston, are examples of this. The tower mill at East Knoyle, however, was externally more reminiscent of her Somerset neighbours, with cloth set sails and a boarded cap turned to the wind manually by winding gear housed in a tailbox. Brinkworth Mill, new-built in 1832, was a six-sailed mill with a dome-shaped cap, winded by fantail, and had auxiliary power by 1871. By 1880 the mill was converted permanently to being engine-driven, the conversion being carried out by Edwin Ratcliffe of the Westport Iron Works, Malmesbury, which works also produced and maintained gear in many of the local watermills and frequently 'dressed and hardened' mill 'pecks'[25] — steel mill bills used in millstone dressing.

Millwrights had become few in Wiltshire by the end of the 19th century, so much so that in 1885 "two Millwrights, well accustomed to Windmill work" were advertised for in a national trade journal to erect the last windmill to be built in the county at Stert, east of Devizes.[26] The situation proved to be similar when the restoration of Wilton Mill was undertaken, millwrights from Warwickshire and Suffolk being brought in to do the work.

The fact that no Wiltshire windmill worked traditionally after the first World War has meant that by the 1970s much of the knowledge and traditions of windmilling in the county have been lost, along with the millers and millwrights. It is fortunate that there are still a few watermills in traditional, if not full time, employment and of credit that Wilton Mill has been restored as a working monument to this ancient tradition.

Winterslow post mill, a drawing after the engraved horn beaker of 1816 (Devizes Museum).

Some Medieval References to Windmills

Some medieval references to windmills of which nothing further is known are listed below:

ALVEDISTON	1272, 1331
WINTERBOURNE BASSETT	1281
BROAD BLUNSDON (Blunsdon St Andrew)	1284
WROUGHTON, Elcombe	1287, 1310, 1348
POULTON (now Gloucestershire)	1297
OAKSEY	1299
BERWICK ST JAMES	1307
YATESBURY	1309
WANBOROUGH	1310
GRITTLETON	1330s
HILMARTON	1348
STANTON ST QUINTIN	1361
STEEPLE ASHTON	1371

The following are possibly sites of medieval windmills:

ALL CANNINGS, Windmill Ball (15c)	SU 074 613
AVEBURY, 1 Windmill Boll (1763)	SU 104 693
2 Overton, Mill Field	SU 117 679
BERWICK ST LEONARD, Windmill Field (1840)	ST 92–33–
HANKERTON, Wind Mill Hill (1809)	ST 962 916
HEDDINGTON, Upper & Lower Millway (1840)	SU 005 663
HINDON, Windmill House (1773)	ST 912 338
GREAT HINTON, Windmill Fields (16c)	ST 90–59–
HULLAVINGTON, Windmill Hill (1599)	ST 888 826
LYDIARD TREGOZE, Windmill Hill (OS)	SU 103 837
MINETY, Windmill House (1808)	SU 014 911
PURTON, Windmill Hill (1515)	SU 08–87–
WILCOT, Windmill Hill (1748)	SU 140 609

*Cornbrash Mill,
Kemble, in 1907
(J. K. Major).*

*Tilshead post mill,
c. 1905
(WA & NHS).*

Aldbourne Mill, c. 1900 (Aldbourne Photographic Club).

ALDBOURNE
SU 270 761

A windmill worth 20s. a year is recorded in 1311 and 1347.[27] The last windmill to stand in Aldbourne was a brick tower mill with a dome-shaped cap and four shuttered sails winded by a fantail. It was standing in 1837, then occupied by Owen Lassam;[28] George New was miller in 1875, Henry New the miller in 1885. The mill worked up to the end of the 19th century and when it was pulled down in 1900 the bricks were re-used to rebuild Windmill Cottages.[29]

ALDERBURY
SU 185 274

Greenwood's map of 1820 shows a windmill in a field known as Windmill Hill close.

ALDERBURY,
Whaddon
SU 196 267

A windmill erected in 1799 is shown on Greenwood's map of 1820 as standing on Whaddon Common. In 1800 it was for sale, containing two pairs of French stones and a bolting mill.[30] In 1806 Abraham Golding, late of Whaddon, miller, was a prisoner for debt in Fisherton Anger gaol.[31] This windmill is not recorded later than Greenwood's map.

ALLINGTON,
Boscombe
SU 197 385

A windmill "timber and tiled including machinery" was insured for £300 by Thomas Waters in 1793.[32] Boscombe Windmill, standing on common ground, is illustrated as a post mill with a roundhouse on the Tithe Map, but is not recorded after the mid 19th century.

ASHTON KEYNES
SU 058 938

Andrews & Dury's map shows a post mill, which was noted as being newly-erected by Maurice Maskelyne Bennett sometime shortly after 1758. In 1778 Bennett was to provide protection for travellers on the new public road from his windmill.[33] It is last shown on Greenwood's map of 1820.

BISHOP'S
CANNINGS
1   SU 024 647
2   SU 038 630
3   SU 057 630

The first site is given as "Windmill Knowl" on the early OS and on Greenwood's map, and is probably a medieval site. Greenwood also shows "Windmill House" and a symbol at the second location and this windmill, which stood on high ground called "Windmill Ground" and a third mill close to the parish boundary, are marked on the 2″ OS preparation maps.

CHITTERNE,
All Saints
ST 999 443

A windmill is recorded on the manor in c.1323, on a site known as Windmill Down.[34] William Watts of Great Cheverell, millwright, insured his "Corn windmill house timber and thatched" for £300 in 1798; the tenant was George Baker.[35] The mill was then newly built and contained one pair

Chute Mill, from an old postcard, c.1906.

of stones for grist and a pair of French stones. It was probably similar in appearance to the post mill at Tilshead. In 1799 the mill was advertised for sale with a storehouse,[36] and in 1831 it was again offered for sale by auction, then standing in good repair, with a new-built house in the village of Chitterne St. Mary. John Cleaver was the tenant, the premises having been lately in the occupation of John Grant, deceased.[37] In 1855 the farmer and miller was John Titt, the father of John Wallis Titt. George Abery, junior, is the last recorded miller at Chitterne in 1859.

CHUTE
SU 296 537

A windmill worth 6s. 8d. yearly stood on the manor in 1331.[38] Andrews & Dury's map marks a post mill "Chute Wind Mill" in 1773 and in 1781 John Hold insured a windmill and "going geers" there for £130.[39] In 1786 the mill is recorded as having a bolting mill[40] and was advertised for sale as part of the manor in 1809 and 1810.[41] In 1840 and 1848 it was owned and milled by Richard Batchelor; Charles Taylor was miller in 1885, Edward Taylor in 1889 and the mill is shown as disused on the 1899 OS map. An old postcard shows Chute Mill as an octagonal smock mill with four common sails, in an advanced stage of decay with props supporting the timber-framed tower. No other smock mills are specifically known in Wiltshire but the description of a windmill "covered with boards" — as at Collingbourne Kingston — could well apply to a smock mill as against a post mill, particularly in the east of the county.

CLARENDON PARK
SU 169 302

Greenwood's map of 1820 marks a windmill north east of Ranger's Lodge Farm.

COLLINGBOURNE
DUCIS
SU 244 534

A windmill on the manor of Collingbourne was worth 40s. in 1361 – 2.[42] Andrews & Dury's map shows "Old Wind Mill Hill" at SU 253 532, and a tower mill symbol at the given reference, presumably the windmill erected in 1767 by John Noyes, miller, on land belonging to the Earl of Aylesbury. After Noyes' death in 1797 the windmill was leased several times and was milled by William Hunt in 1845.[43] In 1855 and 1859 it was held by George Hooper, miller and maltster, but is not recorded in use later. This is possibly the unprovenanced windmill "of seventy feet sail, good width; one pair of French and one pair of Welch (sic) stones; two flour machines, one smutter . . ." which a Mr Hunt of Calne was offering to let in 1857.[44]

COLLINGBOURNE
KINGSTON
SU 234 553

Andrews & Dury's map indicates a tower mill here in 1773 and the 2″ OS preparation survey shows two windmill symbols side by side. It is possible that this is an example of one mill being superseded by a new one, for a reference to a lease of 1820 recounts that Robert Mackrell, farmer and miller, "had erected a windmill" which was left to his wife Elizabeth in 1838, and leased by her to Thomas Wheeler in the same year.[45] In 1839 – 40 the windmill was insured as "a Wind Corn Mill . . . covered with Boards no Oats shelled or kiln therein . . . shaft veins break wheel and 3 large wheels 3 pair of Millstones wire machines & dressing mills & other machinery therein . . . Stable only nr. used as a Storehouse (kiln therein for drying Wheat only) . . ." Thomas Wheeler was then described as a miller and baker.[46] On the death of Elizabeth Mackrell in 1865 the mill passed to Henry Berry, a maltster of Ludgershall, and W. Batten was miller in 1867. The windmill apparently was disused by 1875 but the site can still be located by a mound some 4' − 0" high, now housing a water storage cistern.

A public house by the main road through the village is named 'The Windmill', but its sign depicts a Dutch tower mill!

CRICKLADE
SU 085 934

A windmill, shown as a post mill on Andrews & Dury's map of 1773, stood on the highest point of Common Hill, formerly known as Windmill Hill.

DOWNTON
SU 190 216

A windmill is marked on the 1st Edition 1" OS map, standing to the north side of the road at the west end of Packham Common.

EVERLEIGH
SU 195 538

Everleigh windmill is shown on the early OS and Greenwood's map of 1820. It was still standing in 1843 when it was occupied by George Hooper.[47] By 1900 the site was marked only by a widening of the footpath and nearby 'Windmill Cottages'.

IMBER
ST 962 487

A tower mill is shown, standing above the village, on the Tithe Map of 1838, when it was owned and occupied by William Fricker. James Cruse was miller and carrier in 1848, Alfred Cruse the miller in 1855 and from 1859 to 1885 Edward Hayter is recorded as miller. E. H. Deane was miller from 1889 to 1895 and Thomas Goddard worked the mill from about 1898 until it ceased work sometime between 1903 and 1910. The mill stood until c.1914[48] but the village was taken over by the Ministry of Defence and completely evacuated early in the second World War; the site has not been examined.

KEMBLE
ST 986 977

Although part of Gloucestershire from 1896, Kemble mill was built and worked as a Wiltshire windmill. Ogilby shows a mill here in c.1675, as do Andrews & Dury — "Old Wind Mill" and a tower mill symbol — and Greenwood in 1820. A stone tower mill, called 'Cornbrash Mill' was photographed in a very derelict state in 1907 and the brittle nature of the local oolitic stone — known as Cornbrash — probably accounted for the demise of this mill. J. Hiscock was miller in 1855 and 1859, David Hiscock in 1867 and 1875.

**KINGTON
ST. MICHAEL
ST 90 – 77 –**

A windmill was built here in c.1807 by Thomas Berry and held in trust for his son, Joseph, in 1812. In 1818 "all that Freehold Windmill (except the internal part) with two cottages . . ." was sold at auction to John Woodman, a mason of Chippenham. Joseph Berry was miller at Stamford Mill, Faringdon, Berkshire, that same year.[49] Still standing in 1821 the windmill is not recorded later and is not marked on any of the county maps; its exact location is at present unknown.

**LUDGERSHALL
1   SU 266 509
2   SU 262 505**

Andrews & Dury's map of 1773 shows a post mill at the first site, south of the castle. It is also marked on the 1st edition OS and Greenwood's map. It is said that a neighbour had a quarrel with the windmiller here and planted a belt of elms to the windward side of the mill, which still stood in the 1880s although the mill had gone.[50]
The second site, south west of the church, was of a windmill owned and occupied by James Hunt in the 1830s and run by members of the Hunt family until at least 1875. This site is now obscured by an army barracks.

**MILTON
LILBOURNE
1   SU 188 618
2   SU 181 616**

"Windmill Ground", north west of Totteridge Farm, appears in the Tithe Award. A windmill, shown at the second location on the early OS maps, is mentioned further in 1828 as a "most substantial brick-built windmill" with one pair of capital French stones; the shaft, cogs, sack tackle and dressing machine were in the best preservation.[51] This mill is marked as "New Mill" on Greenwood's map of 1820.

**ORCHESTON
ST. MARY
ST 05 – 45 –**

A windmill at a rental of 10s. a year stood on the estate of Giles Tooker in 1627.

**PEWSEY
1   SU 148 611
2   SU 168 616**

"Wind Mill Ball" is recorded in the Tithe Award close to the parish boundary with Wilcot and a windmill is marked on Greenwood's map of 1820 just south of the Kennet & Avon canal, to the east side of the parish.

**PRESHUTE
SU 167 697**

A timber windmill with two very large Peak stones was to let by Richard Dangerfield, baker, of Marlborough in 1731. It was sited in Manton Field.[53] The site is given as Windmill Edge Field in the Tithe Award, but the mill was not standing at that time and it is not recorded on any of the county maps.

**SHREWTON
SU 072 437**

Shrewton windmill stood on an area known as Windmill Hill, and a Windmill Lynch there is referred to as early as the 1630s.[54] In 1793 William Watts of Great Cheverell insured a windmill "timber built and thatched" for £250, which stood in Mill Field or Nett Field alias Netton,[55] and the mill is described as having a pair of stones for grist and a pair of French stones in 1799.[56] The mill was then rented by George Maslen. It is recorded as disused on the 1886 25" OS map, is not indicated on the 1901 2nd edition and nothing survives on site.

SHREWTON,
Maddington
SU 066 444

The second Shrewton windmill stood about half a mile north west of the previous site and was apparently raised on a Saxon burial mound.[57] In his will, proved in 1662, Thomas Merriwether of Bathampton in Wylye parish left to his son, Thomas, his windmill in Maddington.[58] A windmill is shown on the early OS map and Greenwood's map of 1820. It is marked on the 1901 25" OS map, apparently still in use, but had gone by 1924.

Some confusion exists over the respective windmillers at Shrewton, for no distinction is made between the two mills in the trade directories. It seems likely that the Shrewton mill was held by the Maslen family from at least 1848 to 1875, when James Maslen was miller and baker there; he was, however, declared bankrupt in 1877.[59] W. Curtis was miller at Shrewton in 1853, P. Hayter in 1859 and 1867 and George Williams in 1898. These millers were presumably connected with the Maddington mill.

STERT
SU 032 594

In 1884 an advertisement appeared in *The Miller* for the "Entire Machinery for a Tower Wind Mill, two pairs of stones" wanted by a miller at Stert. The following year two millwrights "well accustomed to windmill work" were similarly advertised for,[60] and the windmill is marked as a corn mill on the large scale OS maps of that time. By 1922 it had gone, the area being allotment gardens. There is now no evidence of its existence and it must have been the last built and one of the most short-lived windmills in the county.

STRATTON
ST. MARGARET,
Kingsdown
SU 172 884

Property bought by Adam de Stratton in 1261 included a windmill, which is further mentioned in manorial court proceedings in 1281.[61] A windmill is marked on the 1st edition 1" OS map, and was part of Kingsdown Farm when the property was let in 1829.[62] In 1859 S. Jones was the miller; by 1900 the mill had gone.

SWINDON
1 SU 157 838
2 SU 13 – 83 –

A windmill on the manor was held by Aymer de Valence in the early 14th century[63] and Windmill Post occurs in 1592.[64] A Windmill Furlong in Swindon west field was named in the 17th and 18th centuries.[65] Wood Street, Old Town, was formerly known as Windmill Street and the last windmill there stood on or behind the site of the present King's Arms

Hotel. It was pulled down by 1867 and the stone used to build three cottages.[66] A windmill in Swindon Field "recently built in a most substantial manner, and fitted with patent sails and machinery of the newest and best construction, having two pairs of wheat stones, and one pair of barley stones, store rooms, bins, &c." with dwelling house, stable, cart-shed, piggery and other buildings adjoining, was for sale in 1852.[67] This was perhaps the windmill built c.1848 – 9 by Isaac Holdway on part of Okus Farm and known as the 'New Mill'. It was destroyed by fire shortly before 1854 and the actual site is unknown.[68]

SWINDON,
Moredon
SU 128 875

"Ridgeway Mill" is shown on Greenwood's map and the early OS maps similarly locate it. The land was occupied by William Hiscocks in the 1840s and on the 1883 25″ map it is shown by a circle marked "Old Windmill".

NORTH
TIDWORTH
SU 247 510

A windmill at Tudeworth was worth 20s. a year in 1290[69] and an indistinct symbol appears on Saxton's map of 1576 on Windmillhill Down. On this hill a sail-less tower is shown on Andrews & Dury's map of 1773, close to the parish boundary with Ludgershall. The area is still known as Windmillhill Down.

TILSHEAD
SU 032 477

William Cooper of Tilshead insured his windmill there for £200 in 1785; it was then in the tenure of Henry Baker.[70] Tithes in the parish were extinguished by an act passed 1811 – 12, "save and except the tithes of a certain mill called or known by the name of Tilshead Wind Mill". Arabella Hussey was the owner and occupier in 1853 and is recorded as miller in 1855. E. Mead was miller in 1859 and the mill was last worked by Thomas Long in the 1880s, who ground flour for use in his own bakery.[71] The mill was sadly decayed in the early 1900s and when it was taken down, c.1905, the timbers were reused to repair the chancel roof of the village church.[72] Tilshead Mill was a post mill with a vertically boarded roundhouse and a thatched roof and must have been similar to the other 'boarded and thatched' windmills on Salisbury Plain. She carried four shuttered sails, with stocks morticed through a timber windshaft and was apparently winded by tailpole. A small flint and brick under thatch store stood nearby.[73] The area, above and to the south of the village, is now allotments.

WESTBURY
ST 870 503

A windmill shown on the 1st edition 1″ OS map is marked as a tower mill on Greenwood's map of 1820, but does not appear later.

WINTERSLOW
SU 249 329

A windmill worth 16s. a year was recorded at Winterslow in 1274. By 1297 it was worth only 10s., but worth 20s. in 1334.[74] Robert Horne owned and occupied a windmill with two pairs of stones, house and bakehouse adjoining, which were for sale by tender in 1810,[75] and the windmill is shown on the early OS and Greenwood's map. A carved horn beaker in Devizes Museum commemorates an extraordinary event connected with this mill. On 20 October 1816 the Exeter Mail coach was attacked by a lioness which had escaped from a menagerie on its way to Salisbury. It was reported that the lioness was pursued "into a hovel under a granary" and, as the windmill, shown as a post mill with a round house, appears in the background of the scene, it was perhaps into the round house store of the mill that the lioness was chased.[76] The mill was occupied by John Judd in 1840.

WOOTTON
BASSETT
1   SU 058 819
2   SU 073 821
3   SU 064 827

On the manor of Wootton there were a watermill and two windmills in 1271; a horse mill is also referred to ten years later. When Vastern and Wootton became separate manors in the early 1330s, Vastern held the watermill, "beyond the reprise" but valued at 13s. 4d. in 1334 and one windmill worth 15s. a year.[77] A windmill associated with Hunt's Mill (water) is referred to in a lease of 1674 and stood on a hill to the north of the watermill. It is shown as a post mill on Andrews & Dury's map and blew down on a Sunday in c.1781. Another windmill stood on Brynard's Hill, on a mound there, and a third stood near the bottom of Wood Street.[78] This latter was standing in 1728 and is shown as a tower mill on Andrews & Dury's map, but does not appear on later maps.

WOOTTON RIVERS
SU 194 627

Andrews & Dury's map of 1773 shows a post mill on part of the Manor Farm lands. No other references have been found.

Bradford on Avon, windmill tower, 1974.

Brinkworth, mill tower and outbuildings, 1929 (SPAB).

Tower Mill Remains

BRADFORD ON AVON ST 828 612

In 1809 Thomas Smart of Bradford advertised for a wind-miller,[79] but by 1817 "all that erection or building, formerly used as a Wind Mill but now converted in to Tenements", in the occupation of Richard Pearce-Ackerman and Thomas Smart, baker, was to be sold by auction.[80] The property was still known as the Windmill or Old Mill in the 1860s[81] and the tower, of well-coursed, dressed Bath stone, with ashlar string courses and a conical slated roof with overhanging eaves, still stands as a private dwelling house off Mason's Lane. It forms a prominent landmark to the north of the town, overlooking the town bridge and river Avon.

BRINKWORTH
1 SU 017 856
2 SU 031 853

Windmillfield and Windmill Leaze are recorded in Brinkworth in 1636 and 1640[82] and similar field names in the Tithe Award suggest early windmill sites to the south west of the village, just north of Brinkworth Brook. Andrews & Dury mark "Wind Mill House" at reference 1., on high ground about 1 mile west of the last site to be occupied in this parish, and in 1771 land called Windmill Grounds at Brinkworth was sold to Young by Haggard and Waldron.[83] This was perhaps the ground on which the last Brinkworth windmill was built in 1832, for it was still owned and occupied by one Abraham Young in 1840, and was built for him.[84] The tower, of soft local stone, was hung with split stone slates and the mill had a high dome-shaped cap, six sails[85] and was winded by fantail. In 1832 it contained two pairs of stones, a dressing machine and had Patent sails, and was built on "the most approved plan".[86] Abraham Young was still miller there in 1848, followed by David Greenman, 1855, Benjamin Barnes, 1859, W. Matthews, 1867, and Henry Hitchcock from about 1870. By 1871 there was auxiliary steam power to drive the mill[87] and a drawing of 9 October 1880, from Ratcliffe's Westport Iron Works, Malmesbury, shows the mill tower converted from windpower to two pairs of stones and a sack hoist driven from an external engine.[88] The two pairs of stones were underdriven by mortice cogged bevel gears. By 1900 a group of outbuildings surrounded the tower, built in local stone and brickwork, and the mill was still grinding, powered by a gas engine, in 1929.[89] It apparently stopped work during the 1930s and the tower, still standing to full height with its cap in 1952 had been truncated and was considerably decayed in 1965

Chiseldon, from the south west, 1974.

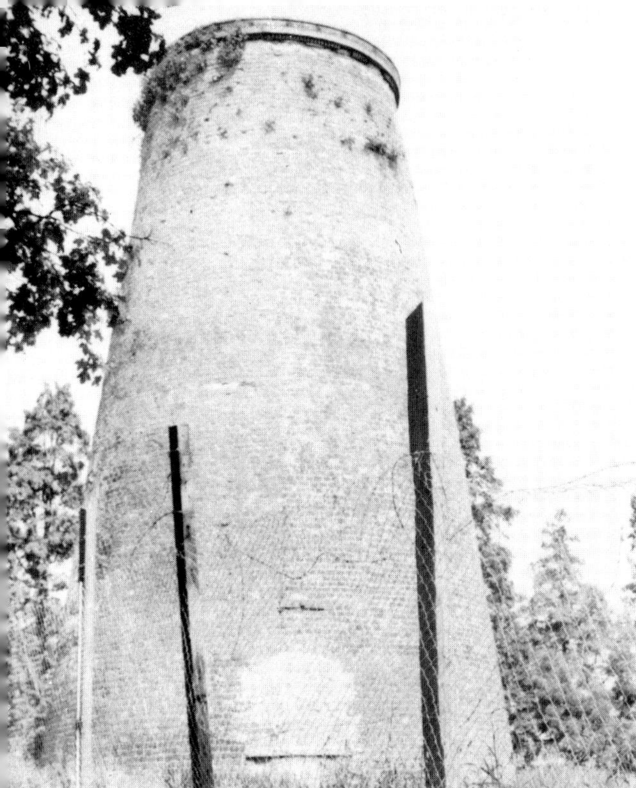

Brinkworth, remains of outbuildings and overgrown tower base in 1974.

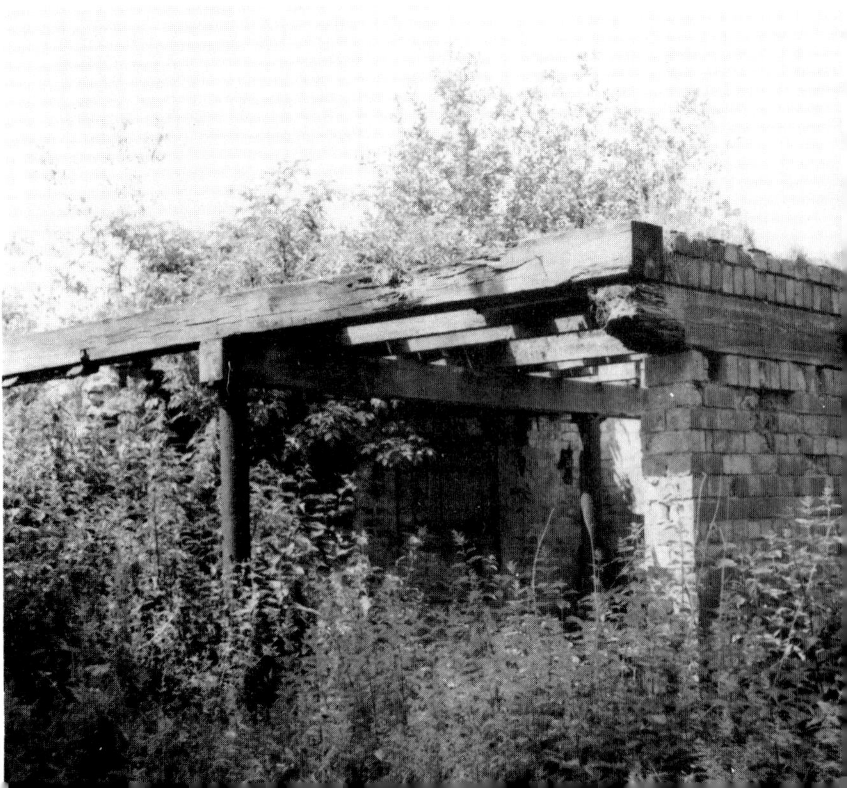

when a survey was made.[90] The remains of the Ratcliffe gear were still *in situ*, steel shafting and cast iron gearing and two pairs of $4'-0''$ diameter millstones, one pair being composition stones by Barron & Son of Gloucester. A single, decayed stone still survives on site. Since 1970 the tower has been reduced almost to ground level and is much overgrown; part of the tower wall, $2'-0''$ thick, stands to a height of about $9'-0''$ on the north side, also the walls of some brick outbuildings and one or two cast iron columns.

CHISELDON
SU 190 799

A "well-established windmill and cottage" was advertised for sale in Badbury in 1859[91] but as Badbury is a tithing north east of the village of Chiseldon (SU 19 – 80 –) it is suspected that this was a windmill other than that of which the tower survives, but further evidence has not been found. The complete brick shell of a windmill tower, $22'-6''$ in diameter at ground level externally, still stands above and east of the church. The tower contained four floors and had a stage at first floor level supported on timber brackets. All door and window openings are now blocked up, the only evidence of machinery being an iron toothed ring for winding the cap on the outside of the curb. The mill was built sometime in the 1820s and the sails are said to have been removed in the 1880s.[92] Part of the inner gear remained in 1913, when it was a storehouse.[93] Young Holister is recorded as the miller in 1848 and John Alder, 'Patriot's Arms' and miller, in 1875.

DEVIZES
SU 001 614

Two brick tower mills were built on the castle mound in the early 18th century for crushing rape seed for oil, the earliest reference being when a half share of the mills was conveyed to William Maple in 1716.[94] A drawing of Devizes from near the North Gate made by William Stukeley in 1723 shows two four-sailed tower mills with rounded caps and tailboxes,[95] while Dore's plan of 1759 shows the south mill with six sails.[96] An 18th century watercolour by Thomas Baskeville shows one of the mills as a four-sailed tower mill with a gable-shaped cap and a gablet extending from the rear, presumably housing the winding gear.[97] By 1740 the mills were used for grinding snuff and a marginal comment on Dore's map notes that "the tobacco trade has grown of late". They were then in the ownership of John Anstie and William Leach, later of William Leach alone[98] and a fine engraved trade card which shows the windmills survives from his ownership.[99] The mills later passed into the possession of William Ludlow, who was still grinding snuff there in 1784.[100] Ludlow subsequently moved his tobacco and snuff trade to Bristol, nearer the supply source of his raw materials. It is said that the mills were converted for corn milling but in 1838 Valentine Leach purchased the castle estate and subsequently built the new

Devizes, remains of brickwork of north tower, 1978.

castle. The south tower was demolished for the new building but the north tower survives, 'gothicised' in c. 1840 to become part of Leach's castle.[101] The brickwork part of the surviving tower is about 12' – 0" internal diameter, but has been much altered internally as part of a private dwelling.

<table>
<tr><td>IDMISTON
SU 177 359</td><td>

Andrews & Dury's map of 1773 marks "Wind Mill Post" in the north of the parish at approx. SU 203 375 of which nothing further is known, and the shell of a tower mill still stands at West Gomeldon, close to the parish boundary with Winterbourne. This mill is said to have been built originally at Middle Wallop, Hampshire, and to have been moved on farm wagons to its present site in about 1840.[102] This removal was presumably only the machinery, for the tower is of local flint and brick construction. The tower, in the form of a truncated cone, is 18' – 0" internal diameter at ground level, with a 2' – 6" thick wall; it is rendered externally and formerly contained four floors, with the stone floor at first floor level. There is no evidence of a stage. The mill was in use until about 1885, latterly milling barley, and a Mr Bray worked there until a week before it closed.[103] All the internal timberwork has now gone, the only remains being some pieces of Peak and French burr millstones and the cast iron windshaft, with its integral canister to carry four sails, the six arms of the iron brake wheel and part of the fantail gearing and support timbers high up and now much overgrown.

</td></tr>
<tr><td>EAST KNOYLE
ST 874 310</td><td>

This steeply tapered stone tower, raised on a circular earth and masonry base some 3' – 0" high and 52' – 0" diameter, was converted into a summer residence sometime after 1912. It stands on a site called Windmill Hill as early as 1632,[104] but is not indicated on maps until the early 19th century. In 1839 it was owned and occupied by Henry Seymour and William Perry was miller there in 1848 and 1859. By the end of the 19th century only one pair of common sails remained,[105] and the mill had a gable-shaped boarded cap with a tailbox containing the winding gear, similar to Somerset tower mills. By 1900 it had ceased work and was burnt out in 1912.[106] The cap was superseded by a four-gabled, tiled roof, which sits on the original curb.

</td></tr>
<tr><td>WINTERBOURNE
MONKTON
SU 105 725</td><td>

There is a long history of windmilling in this parish. A windmill at Winterbourne Monkton is mentioned in the Great Chartulary of Glastonbury Abbey in c.1265 and valued at 40s. in the 1330s.[107] "Wyndemullehulle" is mentioned in 1334 and so recorded until the present day, rather more famous for its archaeological associations than its medieval windmill. It is on the parish boundary with Avebury, SU 086 714.

</td></tr>
</table>

Andrews & Dury's map of 1773 shows "Mill Barrow" also, west of the church, SU 094 721.

Monkton windmill was advertised for sale in 1813 with two pairs of stones, new small house, etc., then in the occupation of William Tully.[108] John Randall was the miller in 1848 and in 1854 he offered the property for sale, described as a "windmill, with patent vanes, driving 3 pairs of stones, iron shafts, capital flour machine, &c, &c, all in excellent repair, with the Dwelling-House and Stabling thereto adjoining . . ."[109] The mill ceased work by 1899. The base of the tower, of coursed roughly squared stone, still stands to a height of about 3' − 0" above an earth mound into which it is built. It is approximately 13' − 0" internal diameter with a 3' − 0" thick wall, and has been lined to form a reservoir for water raised from a nearby well by a Climax wind pump erected sometime after 1924. The wind engine and white-washed mill house form prominent landmarks above the former turnpike road from Swindon to Avebury.

West Gomeldon Mill, Idmiston, in the 1930s (J. K. Major).

West Gomeldon,
remains of windshaft
and brakewheel,
1974.

West Gomeldon,
remains of fantail
gearing, 1974.

Wilton Mill in decay in the 1920s (J. K. Major).

Wilton Mill was built in 1821 by or for John Gale on land leased from the Marquis of Aylesbury. The site of the mill, in part of Forehill Field, had not previously been occupied by a windmill and Wilton Mill was built to compensate for the loss of the Great Bedwyn Town Mill when the water supply was annexed for the Kennet & Avon canal.[110] The mill was apparently commissioned on 1 February 1822, for the leases date from that time.[111] In 1828 the mill was held by William Edwards of Wilton, mealman, but passed to Elisha Edwards on his death. From 1830 to 1853 it was held by Ward, Brown & Co. and milled by George Barnes from at least 1839 until 1848 when his son, John, took over. John Barnes also held the lease of the mill from 1853, and worked the mill until about 1890. The next miller was Thomas Griffiths and in 1906 and 1908 J. Wells is recorded as miller. The mill ceased work about 1914 when it was held by the Potter family. In 1929 the mill and estate were sold to Captain Seminini.[112]

Wilton Mill, built ''upon a modern and most improved principle'',[113] has a tower in the form of a truncated cone of brickwork and originally had a group of low buildings around the base; these were removed in the 1930s. As restored, the mill has a timber stage at first floor level. The cap is dome-shaped and was originally lead covered. This cladding was replaced with iron sheets by White's of Pewsey in 1910[114] and the present cap is of sheet aluminium on the original wrought iron ribs. The whole cap turns on a shot curb with 12 iron rollers and the top of the brickwork is now tied with a concrete ring beam to prevent any distortion of the curb. The cap and sails are turned by a six-bladed fantail on a timber carriage, which is geared to an iron toothed ring on the outside of the curb. In 1978 a hand-crank and disengagement mechanism and a brake were added to the fantail gearing.

At present there is one pair of common, cloth-set sails, and one pair of patent sails, the latter being self-regulating. They are of the type patented by William Cubitt in 1807, with shutters linked together and connected by iron cranks — the spider — at the end of the windshaft to a striking rod and mechanism. By hanging weights on an endless chain which hangs down the lee of the tower from the striking gear, the miller can open, close or regulate the shutters remotely. When the strength of the wind exceeds the effect of the weights, the shutters open automatically and spill the wind. The common sails have to be set individually with canvas by hand. It

Wilton Mill in the early 1930s. (SPAB).

seems that Wilton Mill was built with four patent sails, but this type of sail is less powerful than the common sail and a pair of commons were probably added later to give more power. The roofs of the outbuildings would have provided the miller with a working platform for these sails.

The sails are fixed to a coffin cross cast integrally with the windshaft; the windshaft is circular and tapers towards the tail. The brakewheel, the main driving wheel, is a single casting some $8' - 6''$ diameter, keyed onto the windshaft and carries 108 wooden cogs driven into mortices in the iron rim. These cogs mesh with the wallower, a single cast bevel gear which has 37 teeth, giving an increase in speed of nearly three times. The wallower is mounted at the head of the iron upright shaft and close below it is an inverted bevelled face timber and iron wheel which drives the sack hoist by friction. The sack hoist is engaged by the miller pulling a cord which hangs down through the mill, which raises a similar bevel wheel on the end of the hoist winding drum until it touches the wheel on the revolving upright shaft, turning it by friction between the two wooden surfaces.

Below the cap is the bin floor, through which the upright shaft passes, which has grain bins on one side only. These were for the bulk storage of grain, which is lifted in sacks up through the mill by the hoist.

At stone floor level there are now only two pairs of stones in situ, with one pair set up for milling. These are $47''$ diameter French burrs, enclosed in an octagonal wooden tun with a hopper, fed by chutes from the bins above, supported on a wooden horse. All the stone furniture is new, as is the iron crane used to raise the runner stone for dressing. The second pair of stones are $47''$ diameter Peak stones, for gristing. When advertised in 1828 the mill contained three pairs of stones, a flour machine and a dressing mill. The third pair of stones, located on the south side of the upright shaft, do not have their bridging and support timbers built into the structure integrally, as do the other two pairs, and would therefore seem to be an addition and perhaps the reason that the patent sails were not adequately powerful.

The stones are under-driven, the drive coming from the great spur wheel which is mounted at the foot of the upright shaft, just below stone floor level. The spur wheel has 72 wooden cogs in its iron rim and meshes with the iron stone nuts. These are carried on the stone spindles and can be raised out of gear by a screw and ring device. Each stone nut has 27 teeth. Only the top stone — the runner — revolves and is carried at the head of the stone spindle on a balance rynd or gimbal. The gap between the stones, which must be finely controlled when milling, a process known as tentering, is regulated by a centrifugal governor, having been set before the mill starts work. The governor is driven by chain from a wooden pulley on the

upright shaft just above the spur wheel. Both stone spindles bear on iron bridges hung below the stone floor.

The upright shaft carries an iron mortise crown wheel, with 72 wooden cogs, over the stones. This is an inverted bevel gear and drives a horizontal shaft through a bevel pinion with 28 teeth, which carries a rope-drive pulley at its outer end; from this pulley drive was taken to the floor below to the flour dresser.

The floor at stage level is the Meal Floor and it is here that the meal is collected and bagged or taken to the dresser as it comes down the chute from the stones above. A wire machine, recently brought from a watermill in Wales, has been re-built at this level. The lowest floor of the mill would have served as a store and granary.

Wilton Mill represents much of the refinement brought into English windmilling in the early 19th century; the self-regulation of the fantail, which freed the miller from continually having to watch the wind and trim the mill manually, and the patent sails; cast iron shafting and gearing, usually with the larger, driving wheels having easily replaceable timber cogs; bins built in at the top of the mill, reflecting the change from toll milling, where grain was handled in small, individual parcels, to merchant milling, buying in grain and selling milled products to local bakers and farmers.

The mill was in a very derelict state by the early 1930s and was considered by some to be beyond restoration in the 1960s, most of the timberwork being badly decayed and the iron work in danger of collapse.

In the early 1970s the mill was acquired by Wiltshire County Council and leased to the Wiltshire Historic Buildings Trust who sponsored its restoration to full working order. The millwrighting work was commenced by the Warwickshire millwright Derek Ogden and completed by Messrs Jameson Marshall from Suffolk.

Opposite: Wilton, the cap being removed as repairs begin, early 1970s (WCC).
Overleaf: Wilton, (top left) the brakewheel, wallower and sack hoist, (bottom left) the secondary drive, to the dresser, above the stones, (top right) the great spur wheel and drive to the stones, (bottom right) the stone nut and bridge for the Peak stones.

NO
SMOKING

Wilton Mill restored (WCC).

Notes and References

The only material to be specifically concerned with Wiltshire windmills which has appeared in print is *Wilton Windmill and others in Wiltshire,* D. A. E. Cross, Wilts. Arch. Soc. Bulletin, April 1972, 14 – 16, my own section on Wind Power in *A Guide to the Industrial Archaeology of Wiltshire,* ed. M. C. Corfield, Wilts. County Council, 1978, 27 – 8 and *Wilton Windmill,* Wiltshire County Council and Wilton Windmill Society, 1979.

The basis of this present research has been the material deposited in the County Record Office and various libraries. The following sources have been generally used:

The Victoria County History of Wiltshire, volumes 1 – 10   (VCH)

Wiltshire Inquisitiones Post Mortem, Henry III — Edward II, 1242 – 1326, British Record Society, 1908.   (IPM Hen III)

Abstracts of Inquisitions Post Mortem, Edward III, 1327 – 1377, British Record Society, 1914.   (IPM Ed III)

Abstracts of Wiltshire Inquisitiones Post Mortem, Charles I, British Record Society, 1893.   (IPM Chas I)

The Place Names of Wiltshire, J. E. B. Gover, Allen Mawer & F. M. Stenton, Cambridge University Press, 1939   (PNW)

Publications of the Wiltshire Archaeological & Natural History Society Records Branch, later the Wiltshire Record Society   (WRS)

The Wiltshire Archaeological & Natural History Society Magazine   (WAM)

Wiltshire Tithe Awards, Apportionments & Maps   (WRO/TA)

The Salisbury & Winchester Journal   (SWJ)

The Devizes & Wiltshire Gazette   (DWG)

and the Trade Directories of Pigot, 1830 & 1842, the Post Office, 1848, 1855 & 1859, and Kelly, 1867, 1875, 1880, 1885, 1898, 1903 & 1910.

1   *The Domesday Geography of South West England,* ed. H. C. Darby & R. Welldon-Finn, 1967, 45.

2   *Wiltshire and Somerset Woollen Mills,* K. Rogers, 1976.

3   IPM Hen. III, 3.

4   WRS Vol. XXIV, 1970, & Vol. XIV, 1958.

5   IPM Ed. III, 292, 214.

6   WAM Vol. LXII, 1967, 79ff.

7   WRS Vol. XXVIII, 1973, 93.

8   *A Short History of Broadchalke,* H. M. Trethowan, ND, cover.

9   Copy in WRO.

10 IPM Ed. III, 277.

11 *Britannia,* John Ogilby, c.1675.

12 Copy in WRO; see also reduced facsimile, WRS Vol. VIII, 1952.

13 Photocopies in WA&NHS Library, Devizes.

14 Copy in WRO.

15 *Windmills of Somerset and the men who worked them,* Alfred J. Coulthard & Martin Watts, 1978.

16 *A Guide to the Industrial Archaeology of Wiltshire,* ed. M. C. Corfield, WCC, 1978, 53ff.

17 WRO. 682/7, fol. 81.

18 DWG, 10 Sep 1857.

19 *A Guide to the Industrial Archaeology of Wiltshire,* op cit., 29, & *The Windmills of John Wallis Titt,* J. K. Major, 1977.

20 WRO. Wills.

21 SWJ, 12 May 1746.

22 SWJ, 25 Mar 1799.

23 VCH Vol. 10, 49, and correspondence with Mrs M. Waley.

24 *Reading Mercury,* 4 Aug 1828.

25 WRO. 1269/ (Ratcliffe's foundry documents).

26 *The Miller,* 7 Dec 1885. (SC)

27 IPM Hen. III, 383; IPM Ed. III, 181.

28 WRO/TA.

29 *The Aldbourne Chronicle,* M. A. Crane, 1974 and correspondence with M. A. Crane.

30 SWJ, 3 Mar 1800.

31 *London Gazette,* 2 – 5 Aug 1806. (SC)

32 Royal Exchange Fire Insurance Policy, 11 Dec 1793. (SC)

33 WRS Vol. XXV, 1971.

34 PNW, 481.

35 Royal Exchange Fire Insurance Policy, 31 Oct 1798. (SC)

36 SWJ, 25 Mar 1799.

37 SWJ, 3 Jan 1831.

38 IPM Ed. III, 70.

39 Sun Fire Insurance Policy, 27 Nov 1781. (SC)

40 Royal Exchange Fire Insurance Policy, 4 Aug 1786. (SC)

41 *Reading Mercury,* 15 May 1809 & 8 Oct 1810. (SC)

42 IPM Ed. III, 287, 304.

43 WRO. 9/9/206, 233, 238, 252, 258.

44 DWG, 10 Sep 1857.

45 WRO. 9/10/96 – 101.

46 WRO. 682/7, fol. 81.

47 WRO/TA.

48 *Imber Parish Guide,* ND. (WRO)

49 WRO. 1305/134.

50 WAM Vol. XXI, 1884, 324.

51 DWG, 15 May 1828.

52 IPM Chas. I, 39.

53 *Gloucester Journal,* 4 May 1731.

54 WAM, Vol. LIV, 1952, 416 – 428.

55 Royal Exchange Fire Insurance Policy, 24 Jan 1793. (SC)

56 SWJ, 25 Mar 1799.

57 VCH Vol. I, ii, 473.

58 WRO. Wills, Consistory Court of Sarum.

59 *London Gazette,* 1 Jan 1877.

60 *The Miller,* 5 May 1884 & 7 Dec 1885.

61 WRS Vol. XXIV, 1970, & Vol. XIV, 1958.

62 DWG, 19 Feb 1829.

63 VCH Vol. 9, 131 – 2.

64 *Parish Register of Burials.* (SC)

65 VCH Vol. 9, op cit.

66 *Jefferies' Land,* Richard Jefferies, ed. Grace Toplis, 1896, Ch. 3.

67 DWG, 15 Jan 1852.

68 VCH Vol. 9, op cit.

69 IPM Hen. III, 182.

70 Sun Fire Insurance Policy, 25 Jan 1785.

71 Correspondence with Mr W. Blake, Tilshead.

72 WA&NHS Library, Album DD80 & correspondence with the Rev. F. G. Chamberlain.

73 *Victorian and Edwardian Windmills and Watermills,* J. Kenneth Major & Martin Watts, 1977, 120 & 121.

74 IPM Ed. III, 101, & IPM Hen. III, 108, 215.

75 SWJ, Jun 1810.

76 WAM Vol. XXI, 1884, 348 – 50.

77 IPM Hen. III, 64, 135, & IPM Ed. III, 102 – 3.

78 *The Wootton Bassett Almanack . . . History of Wootton Bassett,* W. F. Parsons, Wilts. Tracts 113. (WA&NHS Library, Devizes)

79 SWJ, 24 Apr 1809.

80 *London Gazette,* 3 May 1817. (SC)

81 WRO. *Survey of Bradford,* 1864, & *Trowbridge Advertiser,* 21 Sep 1867.

82 IPM Chas. I, 205, 297.

83 WRO. 137:17, bundle 53, 1771.

84 DWG, 1 Mar 1832.

85 Wilts. Tracts 53, 1867. (WA&NHS Library, Devizes)

86 DWG, 1 Mar 1832.

87 WRO. 1269/

88 Drawing from Ratcliffe's foundry, now in Athelstan Museum, Malmesbury.

89 SPAB Library.

90 Report by R. P. de B. Nicholson; copies in SPAB & WA&NHS Libraries.

91 WRO. Particulars of Estates.

92 SPAB Library.

93 *Villages of the White Horse,* Alfred Williams, 1913, 106.

94 VCH Vol. 10, 245, 247 – 8, 258 & 259.

95 *Devizes Castle . . . ,* E. Herbert Stone, 1920, fpl34.

96 VCH Vol. 10, fp241.

97 WA&NHS Coll., Devizes.

98 VCH Vol. 10, op cit.; *History of Devizes,* James Waylen, 1859, 402.

99 VCH Vol. 4, fp240; *The English Windmill,* Rex Wailes, 1954, pl. xxix.

100 VCH Vol. 10, op cit.

101 *Devizes Castle,* op cit.

102 SPAB Library; report by Capt. J. A. Batten, 1929 – 30.

103 Ibid.

104 PNW, 177.

105 Two old postcards in WA&NHS Library, Devizes.

106 SPAB Library.

107 *Notes & Queries for Somerset & Dorset,* Vol. XXVIII, 1964.

108 SWJ, 1 Mar 1813.

109 DWG, 24 Aug 1854.

110 *WAS Bulletin,* April 1972, 14 – 16.

111 WRO. 9/27/58.

112 WRO. Wilton Account Books up to 1867; trade directories, and *WAS Bulletin,* op cit.

113 *Reading Mercury,* 4 Aug 1828.

114 *WAS Bulletin,* op cit.

For further information on the development and history of English windmills generally, reference should be made to:

Windmills in England, Rex Wailes, 1948 & facsimile reprint, 1975
The English Windmill, Rex Wailes, 1954 & subsequent reprints.

Winterbourne Monkton: the tower base and Climax wind-engine, 1978.

Books on Wiltshire available from Wiltshire Library & Museum Service

CHIPPENHAM: A BIBLIOGRAPHY, 1980
40p (near print)

CORFIELD, M. C. *ed.*
A guide to the industrial archaeology of Wiltshire, 1978
£3.50

GODDARD, E. H.
Wiltshire bibliography, 1929
£5.50

GREEN, R. A. M.
Bibliography of printed works relating to Wiltshire
1920 – 1960, 1975
£7.50

LANSDOWN, M., MARSHMAN, M., ROGERS, K. *eds.*
Trowbridge in pictures 1812 – 1914, 1979
£3.50

THORPE, P.
Moonraker firemen, 1979
£3.50 (near print)

WALL, A. D.
Riot, bastardy and other social problems—the role of constables and J.P.s 1580-1625
(Wiltshire Monographs No. 1), 1980
£2.00 (near print)

WILTON WINDMILL, 1979
60p.

WILTSHIRE BIBLIOGRAPHY 1976, 1979
£2.50 (near print)

All prices are postage extra